Reg Mundy

About Time -

The Situation of Gravity

Final Edition

for

Mr. Savidge

Contents

Preface   5

Part 1   The Gravity Train

1. A Number of Firsts.   10
   The way we were......
2. Gravity - What it ain't.   18
   The effects are all ......
3. Gravity - What it might be.   24
   The vanishing trick ......

Part 2   What's the Matter?

4. Charge of the Light Brigade.   38
   Where did it come from....
5. Inside a Photon.   43
   Anyone for pancakes ......
6. Movement.   46
   Pass the parcel ......
7. Mass Migration.   50
   All together now......

Part 3   Time Out

8. All the Welkin's a Stage.   57
   Bit Players....
9. Blueprints, Plans and Maps.   60
   Time to take your physics......
10. Truth, Dare, or Consequence.   65
    Free Will and Witchcraft ......
11. Breaking the Mould   70
    The Philosopher's Stone......

Part 4   Any Dog

12. The Right Path.   75
    Paved With Good Intentions....
13. Corroborative Evidence.   77
    Some Explanations......
14. When the Big Times Come
    Captious Sands   82
15. Conclusion.
    The First of Many or The Last   87
16. Addendum on "Gravity Waves".
    What did they measure?   *89*
17. Refutations
    Objections to Expansion Theory debunked   94
18. A Simple Experimental Proof...   100

Preface

"It's all nonsense, isn't it?".

This is one of my favourite lines from the TV series "Father Ted", a comedy classic about three Irish priests and their housekeeper, set on a small island off the coast of Ireland. The line is spoken by someone who loses his faith. He suddenly realises that his entire life has been directed by his belief in a vast edifice of philosophy built on the shakiest of foundations by men (in most similar cases, it is usually men, as opposed to women....) who have either misinterpreted or invented facts to support that philosophy. These men are motivated by a desire to aggrandize the edifice either because they themselves sincerely believe in it (the charitable interpretation) or for their own personal advantage (the uncharitable interpretation). In either case, the result is the perpetuation of a self-appointing priesthood who decry non-believers to the extent of public derision (charitable reaction resulting from a feeling of pity for the poor non-believer) or burning at the stake, etc. (uncharitable reaction resulting from the realisation that others might be persuaded to doubt the verity of the true-faith/ cash-cow/ lifestyle-supporter/ charity/ union/ state/ gang/ financial institution/ hospital/ political party/ and so on, you get the idea....).

I picked the "Father Ted" line as an example, not because I wish to disparage any religious faith, but because it is a great illustration of the moment of epiphany which must have been experienced by many a man who was, say, a dedicated communist under Stalin, a believer in Liberty, Equality and Fraternity of the French revolution who found himself supporting Napoleon, or even famous (or infamous) historic figures such as Akbar the Great whose epiphany, like that of so many others, occurred on his deathbed (- better late than never? Depends on your faith....).

Anyway, to come back to the point (did we leave it?), many of us experience several minor revelations in our lives, for example the shock of realising that the big fat Father Christmas in the department store who is promising us a present if we are well-behaved couldn't possibly get down our little chimney, and the Tooth Fairy doesn't always remember to leave money under our pillow when we lose a tooth. On the other hand, imagine being one of the Holy Roman dignitaries with unshakeable faith in an Earth-Centric Universe, peering through Galileo's telescope at the moons circling Jupiter. It is just a question of degree. So, if we are lucky like most people, we go through life experiencing many minor traumas and a few major ones. You might never experience a "revelation" (innocent?) or become inured to them (cynical?). In the past, there seemed to be more innocent people (gullible?) than there are now, at least in the West, but as the average age of the population has increased there has also been a marked increase in cynicism as, like me, most people become more inclined to question things as they grow older. For those of a poetic bent, the great English poet Mathew Arnold put it more eloquently, viz. "The Sea of Faith was once, too, at the full,....".

I recently went on a short sea cruise, during which several entertainments were provided along with the usual singing/dancing evening shows and assorted quizzes and competitions. One entertainment was a series of lectures by various learned individuals on a curious variety of topics. These included a talk by a gentleman I had never heard of, who made a presentation of his theories on the lifestyle and phylogeny of dinosaurs. I had nothing better to do that afternoon, and having had a long-past interest in dinosaurs, I went along to hear him.

The gentleman was Professor Brian Ford, who explained and illustrated his theories on the lifestyle of several well-known dinosaurs including many of the popular monsters. His points were well-made. The current accepted view of dinosaurs shows them leaping about, all ten-tons or so of them, pirouetting like ballet dancers, and generally conforming to the Stephen Spielberg interpretation. Brian Ford provided a different view. Some of these dinosaurs had their eyes on top of their heads like modern crocodiles and alligators, which would seem to indicate a lifestyle based on hunting with stealth in an aquatic environment. Their long tails, surely an unnecessary extravagance for a land animal, were admirably suited to provide propulsion and agility in water, rather than as balancing weights on land for that part of their huge bodies in front of their legs - surely nature would simply have moved their legs forward a bit.....

Then came the part of the presentation which really brought my interest to focus. Brian Ford produced the quotations and comments from his peers in response to his theories. I was astounded by the vilification and vituperation directed at him by the high-priests of the academic establishment of palaeontologists, just for advocating a theory at odds with their view. To any sane person, Brian Ford's view of dinosaurs inhabiting swamps is just as valid as the accepted view of them running about in deserts similar to those found presently in Arizona. The evidence clearly indicates a preference for dinosaurs to be in or near water, most footprints are found in what was mud, and "land" dinosaur remains are often in ancient river courses or silts laid down in lakes. I felt that Brian Ford at least deserved a hearing, and I deplore the attitude of the establishment in deriding him simply for disagreeing with them rather than refuting his theories with logical argument based on demonstrable facts - facts which in my view are as well-interpreted by Brian Ford as by anybody else. He certainly did not deserve the polemic blast he received from his erstwhile colleagues.

Which is probably why I feel impelled to write this book explaining my disenchantment with several systems of belief which I consider erroneous, plus the possibility that my alternative theories are just as good as those of other people no matter how many acolytes they have gathered. After all, my theories are based on the same unsubstantiated ideas and interpretations of facts/evidence, and therefore as likely to be right, as the view accepted by the thousands of adherents and "high-priests" of the current crop of philosophies. I say "my theories" but I am conscious of the old saying "There is nothing new under the Sun!" so I may not be the first person who ever thought some of them up, but they were original for me when I imagined them.

Now, just as Albert Einstein brought the temples of Newtonian physics crashing to the ground, so I throw my straw into the wheel of Einstein's relativity bicycle on the off chance that it catches in the spokes and throws him over the handlebars. If you are prepared to consider an unorthodox view, if you believe that imagination is the food of creative thought, read on, and I'll give you excess of it.

Part 1

The Gravity Train

1.

A Number of Firsts

The way we were......

After childhood, when life is full of certainties for most of us, there follows a period of uncertainty in our first teenage years. A sudden change of perspective, outlook. First real love, first real tears, first real doubts....

For me, I count myself amongst the luckiest of men, despite a very bad start in life when my father was killed in the Second World War. Without a father figure, I grew up to be independent and learnt to think for myself, and naturally collected more than a few bruises along the way.

However, on the plus side, my life covered the period when civilization moved (forward, I hope!) through the times when piston-engined aircraft became immense jet-powered people carriers, when the public telephone box on the corner was replaced by a mobile phone, when television became ubiquitous, when computers evolved from huge valve-powered code breakers to incredibly potent portable data processing and communication devices, medicine advanced by leaps and bounds, and life improved immensely, at least in most countries. Imagination amongst authors and thinkers ran wild, and I wandered across the red sward of Mars with Edgar Rice Burroughs, fought for and against the irresistible pyscho-historic laws of Hari Seldon in Issaac Azimov's empires, was amazed by the Pillars of the Dawn at the centre of the galaxy in Arthur C. Clarke's Childhood's End, blazed across the vault of time knuckles white on the back of Stephen Hawking's wheelchair, and much, much more.

Most important of all, a good education was available for many, and I took advantage of the opportunity. Unfortunately, once again I got off to a really bad start. At three years old, an only child, I was preventing my widowed mother from getting a job, and the only local nursery school was already full. My mother's eldest sister, a formidable woman I loved dearly, marched me down to the nursery, and pointed out in no uncertain terms that the nursery was intended for the likes of me, whose father had been killed in the service of his country and whose mother needed to work, and not for the children currently being given free nursery places so that their mothers could earn pin money - either I got a place or she would raise such a stink the entire country would know how the local bureaucrats and politicians were abusing the system.

Needless to say, I started nursery school immediately. At the time, I had no idea why the adults in charge glowered at me, and seemed to single me out for smacks whenever there was any naughtiness. As a late-comer, the other kids seemed to resent me too, and I soon learnt to become a good fighter.

The relationship with teachers carried on into my infant school, due to the intimate relationship between the school and the nursery, so I continued to get more than my share of thick ears, although I got on much better with my contemporaries and made some lifelong friends. But the consequences of my unusual entry into the world of education still had a sting in their tail. On leaving infant school, and entering the wider world of the junior school, placement in the hierarchy of classes was by recommendation of the infant school teachers. I found myself in class 1D, out of four classes. Being an only child, during bad weather when I stayed in my grandparents house, I occupied my time by reading comics and books.

After languishing bored witless for two terms in class 1D, which was conducted more or less as a mechanism for keeping less-academically-able children off the streets, the whole school year was led out, one class at a time, into the playground where each pupil was summoned in turn to a large desk occupied by the deputy headmaster, a Mr. Tebbutt. It was a hot sunny afternoon which I remember well. Each child was asked a few questions, and required to read a line or two from a book as pointed out by Mr. Tebbutt. Naturally, class 1A was first, and as there were up to 40 kids in each class, by the time he reached me in alphabetical order, he must have been pretty fed up, I think there were only about half a dozen behind me. Anyway, he asked a few questions to which I gave desultory answers, then indicated a simple line in a children's book for me to read, obviously not expecting much response. I read it out. Mr. Tebbutt looked at me in surprise. He indicated another sentence, which I read out. He selected another book, about the level of a "Biggles" story, which again I read out. Finally, he brought forward a reference book, from which I again read out a passage. Mr. Tebbutt summoned Miss Williams, who was acting as his secretary. "Miss Williams, put this boy in 1A" said he, dismissed me with a wave, and moved on to his next victim.

The following morning, after registration, I was taken along to class 1A together with a couple of other pilgrims and given a desk. Simultaneously, a couple of sobbing teachers' pets were removed and cast into the limbo of class 1B. Throughout the school, similar tableaux were taking place as the powers that be reordered their crop for optimum returns. Despite being probably the greatest beneficiary, I was struck at the time by the unfeeling cruelty of the system, but am at a loss to suggest a better or fairer replacement. Are the only choices either to raise false hopes which will be cruelly dashed when the reality of life is experienced, or to stifle any dream of advancement before it takes root? Looking back in later life, I became (and still am) a fervent supporter of examinations conducted by independent evaluators for assessing academic ability, and deplore the stupid system where teachers/lecturers/mentors can propel their favourites forward by recommendation. In the end, it is unfair, unjust and subversive to society, but we are as likely to put and end to it as we are to abolish nepotism, privilege, and the class system.

Enter Mr. Savidge

So, there was I, at the end of my junior years, entering my teens and senior school with a mixed attitude towards teachers, some of whom I despised, and some of whom I revered. I was lucky with my choice of senior school, it was one of the very first comprehensive schools in England, and had a complement of exceptionally good teachers. Unusual amongst them was Mr. Savidge. I later discovered he was actually a Physical Education teacher who, growing too old (I assume, but perhaps for other reasons), converted to teaching Physics - there was a dearth of Physics teachers at the time. He approached his subject in a determined and conscientious way. He was a man who knew his duty, and by God he was going to do it. And he did it, by the book. He introduced me, and dozens of others like me, to the world of

Physics, for which I will always be grateful.

He displayed little or no imagination, always referring any question to the appropriate book on the subject, of which he had an encyclopedic knowledge up to what used to be described as the "Advanced Level" curriculum. He was the God of Physics, just as Eddie Gray (the deputy headmaster) was the God of Maths. I now started to enjoy school, and looked forward to each new revelation in Optics, Magnetism, Electricity, and the rest. At the same time, my taste in literature had changed from an early interest in comics through thrillers and lurid cowboy epics to a more catholic selection.

Previously, my maiden aunt who was working in Canada used to send American comics to me, and I looked forward to the arrival every few weeks of the fat cylinder of rolled-up glossy paper in the post. When I had read them, I used to take them to the local market place where, on one day each week, bookstalls would offer a wide choice of pulp paperbacks. British comics of the time were printed on flimsy matt paper, and my glossy American comics were much in demand by market book traders. I could take my pick of as many paperbacks as I could read in exchange. I would browse the serried ranks of books looking for my favourite authors ranging from Zane Gray through Edgar Rice Burroughs, but pretty soon I had read all of the Tarzan books, and the horse operas seemed to settle to a tired formula where the hero fought manfully against adversity before shooting the baddy in a last-chapter showdown. Then one day I picked up a misplaced book amongst the cowboy sagas, it was "The Metal Eater" by Roy Sheldon (a pseudonym of E. C. Tubb) which should have been in the Sci-Fi section. Despite it being just another cowboy epic set in space, I was fascinated by the new worlds it opened up, and was soon devouring all the Sci-Fi I could find. This tied in very well with my interest in Physics, as it seemed to me that a thorough understanding of Physics was a necessary precursor to any future in space exploration and astronomy.

There was a problem. It seems that good old Isaac Newton hadn't got it quite right. My ultimate being, possessor of the greatest intellect ever to exist (in my opinion), might possibly have feet of clay after all. Some upstart had invented a new set of laws of Physics, and it stated that nothing could go faster than light. There was no way of getting to even the nearest star in anything like a reasonable time, and all these wonderful mechanisms such as hyper-drive, sub-space drive, time dilation, bloater drive, different dimensions, were just crazy ideas that could never work.

I directed my questions about relativity to Mr. Savidge. He could give no answer, only pointing to the relevant books which detailed the new regime without, in my opinion, any solid proof of it's veracity. OK, there were results of experiments such as the discrepancies in planetary movements as predicted by Newtonian Physics which could be explained by the new Einsteinian theories, but there were other equally likely explanations which could be imagined. Mr. Savidge could not help, and we had come to the point where, regretfully, I must leave Mr. Savidge and set off on a new course. My whole universe based on Newton's laws was crumbling, the old certainties were dissolving before my eyes, my faith in lore handed down reverently by distinguished academics was destroyed.

Epiphany. And the Rubicon waited to be crossed.

2.

Gravity - What it ain't.
The effects are all ......

One of the questions which I once addressed to the now-unfortunate Mr. Savidge was about Gravity. "What is it?". The answer, pointed out by Mr. Savidge, was straightforward. It is the attractive force between two masses, proportionate to the total mass and inversely proportionate to the square of the distance between them. Simple.

"No," said I, "that is the EFFECT of Gravity, not what it is!". "Well," countered Mr. Savidge, "it is a force similar to magnetism!".

The discussion raged over a period of time, but in a nutshell, it seems that gravity cannot be deflected, augmented, negated, shielded against, reflected, reproduced without mass, seen, felt, smelt, or otherwise detected. There is no particle associated with it, as in the case of electrical forces (the electron), light (the photon), sound (vibrations in a medium), heat (the radiated sort, carried by wavelengths longer than light in a similar manner). Now, I know that great efforts are being made to detect gravitational waves, find the "gravity particle" (currently thought to be Higg's Boson), and so on. Einstein depicted gravity as a dimple caused by mass in the fabric of the space-time continuum, which to me seemed as likely as hyper-drive (where is the evidence that there is a space-time continuum where mass can cause a dimple?).

I confidently predict that no "gravity wave" from an exploding/imploding galactic object will ever be detected, and that, whilst many more particles may be discovered including the elusive Higg's Boson, they won't have any properties, other than their mass, which provide "gravity".

There are some interesting problems when considering the effects of gravity. Taking two objects, the "gravitic force" between them draws them together with an effect decreasing in proportion to the square of the distance between them. This force slows them down if they are moving apart, starts them moving towards each other if they are stationary relative to each other, or accelerates them towards each other. Although the force reduces dramatically with increasing distance, no matter how far apart the objects, there is SOME force. Eventually, even objects moving apart at almost the speed of light, after an enormous but FINITE length of time, will be brought back together (does this apply to photons, which are moving at the speed of light, and so have infinite mass - but they don't! A bit of a problem for current theories...More on this later!).

Therefore, any two or more objects cannot escape from each other, eventually they must move towards each other or orbit one another. In the most familiar case, where one object goes into orbit around another (actually, both round a common focus(s)), the motion is in the form of an ellipse (it can be a circle if both focal points of the ellipse are in the same place!). Unfortunately, the mathematical formula governing these movements show that, given sufficient relative velocity, the motion of the objects forms a hyperbola (or in a special case, a parabola), and the objects fly apart rather than orbit each other. This would seem to contradict the idea that gravity knows no limit, some force would always be present, and therefore it follows that every object which has mass forms part of the centre of a black hole, from which nothing can escape. The black hole can be almost unimaginably huge, formed say by a photon, or smaller as the mass of the object increases or the objects' relative velocity with regards to each other decreases, but the principle remains.

This brings me to a point where my view of the space-time continuum, the cosmos, or whatever you want to call it (I think I will use the archaic word "Welkin" to distinguish my particular view of the soup we inhabit from the words used by currently popular theories), differs from the accepted norm. It concerns the speed of light, which according to all accepted theories is the same for all observers in our universe. I believe it to be a misnomer. I theorize that, in fact, light itself (i.e. travelling photons) actually move at one zillionth of a percent less than the speed of light, where a zillionth is a decimal point followed eventually by a one but separated from it by innumerable zeros. This seems a contradiction in terms, but if we define light speed as the speed at which an object's mass becomes infinite and which cannot be exceeded in our cosmos, in our welkin it will be found that light itself travels slightly slower.

As further evidence for the variation in the speed of light (radiation) from the maximum possible speed, consider the fact that, if we live in an infinite universe (one with no bounds), in every direction we look there must be at least one star emitting light (radiation). In effect, the whole sky is packed with radiating objects at varying distances with no gaps in between. If light/radiation carries on at light speed forever, every emitting object will contribute some radiation, no matter how reduced by distance, to fall on Earth - there are infinite numbers of objects in every direction, we should be burnt to a cinder! Does this mean that our universe was created at a definite time, and we are only receiving radiation which has been travelling since that creation? In that case, stand by for ever-increasing background radiation levels as our the sphere of contributing objects has a radius increasing at the speed of light, and must eventually become almost infinite!

With our ever-improving telescopes and detectors, are we getting close to seeing the edge of creation, light emitted at the very moment of creation of the universe? In that case, the "edge" of creation could be akin to a flame expanding through a gas cloud, and not the expanding "edge" of a big bang. I'll explain why I think this in Part 4, which covers the theory of what is happening "at the next level down", but it means that our welkin is not expanding like a balloon but simply increasing in size - maybe not really, only our light-limited perception of it! Or, of course, that light does not go on forever, but gets "tired", or wears out, or otherwise changes. Generally, light from further away shows a longer and longer wavelength (red shift) which is commonly ascribed to an expanding universe ----
(see Wiki entry for Edwin Hubble).

If we assume that, in fact, light does change, and the universe is not expanding, we now have a welkin with a fundamental difference to the cosmos described by Newton or Einstein. I forecast that in the near future an experiment will be designed and performed which proves this assertion, and it will be based on the assumptions in the next chapter, which indicate that GRAVITY IS NOT A FORCE and light does not lose energy working against it but for a quite different reason.

3.

Gravity - What it might be.
The vanishing trick ......

When you get right down to it, every time science enables us to increase our powers of magnification or detection, we find whatever we are looking at is actually made from smaller particles than we expected. Thus, when we were eventually able to see molecules, we found they were made of atoms. When we could "see" atoms, they turned out to be a cloud of electrons surrounding a tiny, tiny nucleus. It seems to be inevitable that, when we are able to look at the nucleus and see the individual protons and neutrons, they will turn out to be made of smaller particles again, and so on.

I looked forward eagerly to university, where I expected to be able to debate the fundamental properties and laws of the cosmos with vastly superior minds who would provide dazzling insights to their workings. My university was King's College, University of Durham, based in Newcastle upon Tyne (in the North-East of England, for those who want to find it for any reason). It is now Newcastle University, and separate from Durham. When I applied to go there, it had a famous reputation as a centre for Physics, and I was elated to gain entry. Alas, I was to find that the over-riding interest of the Physics Department was Geophysics, in particular Tectonic Plate theories. My interests were focussed on matters of galactic import, not fiddling about with the thermodynamics of an insignificant ball of rock. My popularity amongst the academic elite soared with my comment that I didn't necessarily believe the surface of the earth was a series of plates, rather it was similar to a bubbling saucepan of porridge which has temporary relatively solid parts on its surface. This was an offhand suggestion, already put forward by someone else (I can't remember who) and I wasn't a serious supporter of it, merely a devil's advocate. Of course, it didn't endear me to the Tectonic Plate Nazis, who returned my attitude with interest if I tried to discuss anything serious with them, such as "What is gravity?".

Eventually, I gave up arguing for my own pet theories and sank into superficial acceptance of the prevailing theories. In lectures, I listened with increasing disquiet to the latest theories of atomic physics, the discovery of ever more and increasingly weird particles which had properties like charge (OK), spin (OK), colour (mmmm?), strangeness (What?), smell (you have to be joking...), and so on. I just couldn't believe that nature (or God if you prefer) would make things so complicated. In my experience, at a fundamental level things tend to be simple, and the most complex structures are actually comprised of different combinations of relatively simple things arranged in a wide variety of ways, like DNA for example. Modern Physics was looking at complicated structures and treating them as fundamental, a bit like having a name for every single building when what you really need is an understanding of bricks, cement, steel, etc., and a plan. Well, I had to accept all this if I wanted to gain the grudging approval of academe so that I could graduate and get a job. But I still harboured my own grand scheme of how it all works, I just didn't try to get other people to listen.

Many years later, along came Brian Ford, and I thought, "Why not? To hell with public derision or approbation, what do I care?", and I therefore lay out my theories for your consideration like so.

Lets assume there is a fundamental particle, right down at the basic level, from which all matter is formed. Inevitably, if we ever got down to being able to "see" it, it would itself be comprised of even smaller fundamental particles, and so on.... But for the moment, in our "macro" world, we can assume it is the smallest bit. We can assume it has certain properties, and that one of these is an electric charge - I make the possibly unwarranted assumption that a "neutral" particle is actually a combination of positive and negative particles, in line with my belief that matter and energy are interchangeable (I have no argument with $E=mc^2$) and the fundamental particle is actually an electric charge. Furthermore, we see from numerous experiments that the smallest element of charge we can identify is not the electron charge but one-third of the electron charge. So, I am going to call my new fundamental particle (or charge) a TIRD, because it is either a third of an electron or less (in reality it may be several smaller bits). The electron charge is 1.6021765 × 10–19 coulomb (see www.britannica.com/EBchecked/topic/183512/electron-charge)and Quarks always have a charge of one third or two thirds of this value. The charge may be positive or negative, so for convenience I will call the positive tird a "t+rd" and the negative tird a "t-rd".

Logically, an electron must consist of at least 3 t-rds, or 5 tirds (4 t-ds and 1 t+rd), or 7, or 9, and so on..... (I subscribe to at least 5 tirds, arranged as four t-rds associated with one t+rd). Similarly, a photon must be 2 tirds (1 t+rd and 1 t-rd), or 4, or 6, and so on.... Now it is commonly held that a photon is massless. At least all experiments so far show that it is either massless or very, very, close to it. I subscribe to the "very, very, close to it" school, as photons travel at the speed of light (almost!). This wisp of mass is inferred from the commonly-accepted equivalence of mass and energy, so I believe the photon is 2 or more tirds which are electric charges, i.e. energy.

So, lets look at what is happening inside the simplest possible model of a photon. There are two tirds attracting each other by virtue of their opposite electrical charge. When two objects attract each other, they can either collide or go into orbit around each other - see Chapter 2 - and given that the mass of the objects is vanishingly small, the attraction of opposite electrical charges must produce correspondingly huge acceleration towards each other and require almost infinite escape velocity to achieve separation.

On the scale we are talking about, the distances between the two objects is tiny whereas the distance to any other objects is immense, so effectively, the two objects form their own "black hole". Now, the objects are falling towards each other and at first sight must either collide or orbit each other. But look closely at the orbit scenario. What happens when one object orbits another, like, say, the Moon going round the Earth. The mutual attraction force is used up/cancelled out by changing the direction of travel of the objects, i.e. a momentum change, so that the constantly-changing path is circular or elliptical. The momentum of an object is a function of its mass and its velocity. But in the case of the two tirds, there is virtually no mass! Therefore no momentum. Therefore the two tirds fall towards each other and there is no alternative.

So they crash into each other? Well, no, they don't. It's like the old saw about the man outrunning the bullet. As the man runs away, the bullet eventually arrives at the point in space where the man was. But he has run further away, although the distance between them is greatly reduced. A period of time passes, and the bullet arrives at the place where the man was previously. Again, he has run away.

This scenario continues, each time the distance between the bullet and the man is decreased, but, no matter how small, there is still some distance. It takes an infinite time for the bullet to reach the man. In our "real" world, of course, the man gets shot, but inside the photon, the rules are different. The tirds rush towards each other at ever accelerating velocity. The end result of this is discussed in Parts 2 and 3, but for the moment we concentrate on this situation inside a building block of our welkin.

Extrapolating this to all the objects in our welkin, electrons, protons, neutrons, etc., everything is comprised of zillions of tirds rushing towards each other. That's EVERYTHING, including you and me (the observers of our universe - sorry, welkin.). All particles/objects are ultimately made from two or more tirds which are getting closer together. Like a house made from bricks when the bricks get smaller, the house gets smaller too. The photons, electrons, protons, neutrons, positrons, bosons, quarks, etc. are shrinking.

On the other hand, larger objects are not. Atoms, for example, are still made from the same number of protons and neutrons as before. This makes for an interesting effect, as, from the point of view of anything made of tirds, the atoms are expanding! It's rather like the Incredible Shrinking Man standing in a room, not realising he is shrinking, the room to him is expanding. To an outside observer, the room stays the same size, but for us in our welkin, made of tirds, the room/atom is expanding. Seeing as we ourselves are expanding as well, we can't see this.

In Part 3 I make my proposition about the quantum nature of time, but for the present, please assume, for the sake of this argument, that time is in fact non-continuous, and consider the following diagram of two bodies orbiting each other in a circular orbit. The two bodies actually move around each other with a central point somewhere on the line between the centres of the bodies, displaced from the midpoint of that line towards the body with greater mass, but we will take the extreme case where the body with greater mass is so much more massive than the smaller body that the focal point is a negligible distance from the centre of the more massive body.

For the sake of simplicity, we will say this is the Earth and the Moon, with the Earth assumed to be so much more massive than the Moon that the Moon orbits a stationary Earth. The V is the velocity of the Moon, and D is the distance between the Earth and Moon as measured by a standard rod of length D.

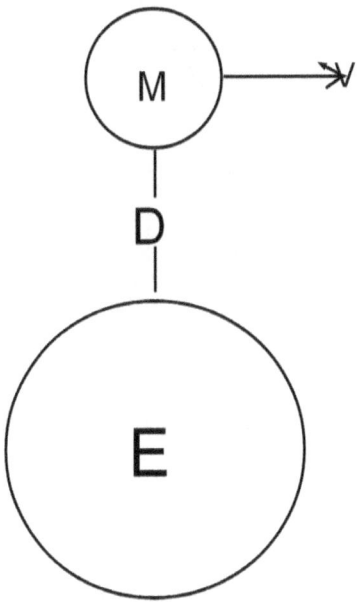

A quantum of time passes, and the diagram looks like this:-

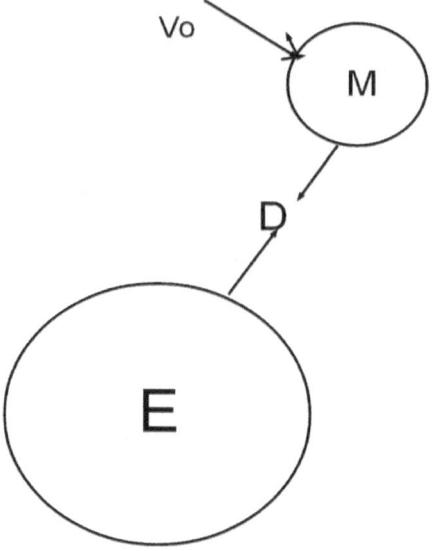

Remember, NO GRAVITY, so the Moon continues serenely on its STRAIGHT path. But wait a second (or a quantum!), the Earth and the Moon are expanding, and so is everything else including the standard rod and the observers (that's you and me!). Everything remains in proportion, the ratio of the radius of the Earth to the radius of the Moon and the length of the standard rod (and the size of you and me!) remains constant. So, we see that, to us, the Earth is the same size, the Moon is the same size, the distance away is the same (or nearly), but the Moon has moved round in a curve! It is travelling in a different direction (the old velocity V is replaced by the apparent new velocity V+).

Another quantum of time passes. To us, the Moon continues in its circular (OK, elliptical!) orbit round the Earth, but in fact the Earth is bigger, the Moon is bigger, and the distance between them is greater. But everything else is bigger, including us, so we can't tell the difference, nothing has changed except the Moon has moved round a bit. (The angle subtended by the Moon to an observer on the Earth is roughly proportionate to the diameter of the Moon divided by the distance - if they both double, the angle remains the same).

So to us (and anything else made of tirds) standing on the surface of the Earth, the swelling Earth beneath us is pushing us outwards, i.e. accelerating us at a perceived rate of 32ft per second per second, or just less than 10 metres per second per second. No gravity, just acceleration.

An interesting speculation arising from this is the relationship between that acceleration (or rate of expansion) and time. Is the acceleration measured in units of time (seconds) or is the time measured in units of acceleration? In other words, is time itself not a standard ever-constant river flowing always forward, or is time simply our way of relating to the current state of inflation of our universe? Our welkin is starting to look more and more interesting, but more about time in Part 3.

Anyway, back to the fundamental question as to why

an object made from tirds should expand, and whether our welkin is expanding as well. If you consider the basic building block to be a tird (even if a tird is made from myriads of other bits), and the basic particle of matter is made from a minimum of two tirds, then if the tirds are massless, the opposing charges of the t+rd and the t-ird will cause them to rush together as there is no mass and therefore no momentum or angular momentum to be preserved - they will simply coalesce.

However, as the tirds are vanishingly small (just a location really, a black hole) the distance between the tirds compared with their size is unimaginably immense. The tirds rush together but take an infinite (or nearly infinite) "time" to collide. Meanwhile, objects built from these shrinking "bricks" are, in comparison with the bricks, expanding. The Incredible Shrinking Man doesn't know he is shrinking while he watches the room grow larger and larger.

You disagree? You think the room is shrinking as well, not growing larger? Well, perhaps it is, but we agree that the size of the room is changing. If the room is shrinking, it merely means that TIME IS RUNNING BACKWARDS. Our time, that is. Current theories agree that time is relative, and depends on where you view it from, and there are many instances where even our local time seems to run backwards, e.g. in atomic physics some reactions occur in a collision of particles where a component of the reaction appears spontaneously in space and joins the reaction, rather than being created by the reaction and then disappearing. How did it know to appear - the reaction did not take place until after its appearance.

You may be astounded that, having demolished gravity, I am now casting doubts regarding the sanctity of time's arrow. For my part, I feel I may as well be hung for a sheep as for a lamb, so I state now that I do not believe time is an ever-flowing-one-way river. More about time in Part 3, but meanwhile if we accept that in any time-frame the size of objects is changing, we don't need gravity.

Maybe, in a few hundred years time, some callow youth will address a re-incarnation of Mr. Savidge with the question "What is gravity?" and he will reply, according to the book, "Gravity was a mythical force invented by the ancients to explain why the Moon goes round the Earth, why the planets orbit the Sun, and so on. The ancients had some weird ideas, they believed the Earth was flat and it was the centre of the Universe! Many people firmly believed in gravity, and a lot of prophets made their careers from supporting it.".

In many a romantic novel, a young couple kiss for the first time under the stars, and in the old cliche "feel the Earth move under their feet". Well, its true, they did. The Earth flung them out towards the stars, and, as it always had, continued to do so for the rest of their lives, unless they became astronauts. We are being accelerated outwards at nearly 10 meters per second per second towards our planetary neighbours, including the Moon, and the nearby stars.

Fortunately, we will never reach them as, whilst they too are expanding, they are also retreating away from us at a proportionate rate, as is the rest of our universe. As the universe is expanding, everything in general is also rushing away from everything else. This starts to break down at astronomical distances as the rate of expansion of the universe moves the edge of our sphere of existence away from us at relative speeds approaching that of light, hence we get a red shift although, relative to us, the universe is not expanding at all. Remember that as light is comprised of photons, it too is "expanding" in its own way. Its actual speed is increasing but remains the same relative to us and everything else in our welkin. In the diagrams above, the measuring rod D can of course be a beam of light.

Part 2

What's The Matter?

4.

Charge of the Light Brigade

Where did it come from.....

The postulation is that all matter is made up from fundamental building blocks consisting of two or more tirds. The tirds are virtually massless, consisting of only a positive or negative charge as far as we are concerned. In reality, each tird is made up from its own building blocks each of which consists of two or more bits on a scale just as microscopic to the tirds as the tirds are to our universe, and those bits consist of .... and so on, ad infinitum.

At our sub-atomic-particle-sub-level we will consider the tirds to be our fundamental particles. So, looking at a photon and assuming it to be as simple as it can be (the welkin might consist of multiple smaller particles than photons, in which case substitute these smaller particles for the photons in this argument), we have a t+rd and a t-rd attracting each other. How did this come about? I subscribe to the theory that, wherever there is a positive charge, there is a negative charge, so, if a positive charge is created there must also be a negative charge created, and vice versa. If you create a positive charge, say in a gas, it will immediately be surrounded by induced negative charges each of which will be surrounded by positive charges and so on, and the same if you create a negative charge. Rather like magnetising a piece of metal, you cannot create a north pole without creating a south pole at the opposite end (forget monopoles, they are a sci-fi invention, have never been found and I believe never will be).

If the accepted view of quantum mechanics and Heisenberg's Uncertainty Principle is applied, wherever the two charges are they can be broken apart by quantum effects and/or other nearby charges and the two charges can go their separate ways. I suggest this is what happens in black holes, where matter is broken down into separate massless tirds which float away, effectively leaving our universe which consists solely of matter - as well as material objects every type of radiation has a particle associated with it which is not completely massless. It follows that our welkin is simply an aspect of an immense sea of restless tirds where a t+rd and a t-rd have occasionally managed to find one another and form a particle. In general, the tirds are the dark matter which astronomers seek, and can never interact with our welkin which only exists as matter, this matter being formed when a t+rd and a t-rd collide in such a manner (influenced by other tirds in the vicinity) that they manage to go into orbit round one another, i.e. their angles of approach and velocity allow them to form an ellipse rather than a hyperbola and they remain in association. Now to do this we have to assume that the tirds do have some mass, however tiny, otherwise there is no momentum to oppose the electric attractive force and they would coalesce (annihilate each other). Their mass is created in exactly the same way on an immeasurably smaller scale as the way they are now going to create mass in the photon they form.

Now we have a t+rd and a t-rd, each of incredibly small mass, approaching each other propelled by their mutual attraction. As they become closer, so the attraction increases and the acceleration mounts until the velocity relative to each other approaches light speed. The relativity effect kicks in, and the mass of each tird increases towards infinity. Unless they are on a perfect collision course (unlikely, given their incredibly small size, the relative distances involved, and the necessary random interaction with other tirds which enabled their mutual capture in the first place), then the now-present mass supplies the momentum to defy their electric charge attraction and cause them to whiz round each other forming the ellipse which is the photon we seek. Of course, the photon may have more then two tirds, in which case we have good grounds to assume the orbits of the tirds round the focii are in more-or-less the same plane, like planets round the Sun, at least for photons.

We now have a photon with mass, created from the energy of the relativistic velocity of the tirds. Crush these photons together, and the tirds either mutually annihilate or wander off separately, and these are the conditions in a black hole where matter leaves our welkin.

Well, we have mentioned black holes, so I suppose we must deal with them. How can you have a black hole without gravity? Surely it is gravity which increases almost infinitely in power as more mass is added to the black hole, until it is so powerful as to drag light itself (almost massless, but not quite) back into its maw. Rubbish! There is no gravity - see Part 1. What actually happens is this. The fundamental building blocks of matter are expanding. The more matter you put together, the more each building block jostles with its neighbours for room to expand, thrusting against them. Given sufficient matter, a sphere is formed. As more matter is added, the sphere grows larger, and the faster its radius must increase to accommodate the ever-expanding building blocks. The faster the radius expands, the faster the surface of the sphere must rise, accelerating at a rate for an object with the same amount of matter as the Earth at 10 metres per second per second. All this acceleration requires energy, and the building blocks are finding it more and more difficult to expand, crushing against each other so that ever more exotic matter is being created by the forced proximity of the building blocks.

Eventually, as more matter is added to the sphere, the surface of the sphere accelerates to the speed of light and the force needed to accelerate further becomes infinite. Light itself cannot outrun the increasing radius of the sphere. The laws of physics demand that the building blocks expand but they cannot, so must break down. Matter leaves our welkin and reverts to free-floating massless tirds.

None of this requires gravity.

5.

Inside a Photon

Anyone for pancakes....

We have assumed that a small particle in our welkin, the photon or another, is comprised of two tirds in orbit around each other. They would follow an elliptic path with the centre of the photon being midway between them if they were of equal energy, but the slightest difference in the energy between them would cause the centre of the photon to oscillate.

If we operate in a three-dimensional welkin, then viewed along one plane we would "see" an ellipse whereas from the other two planes we would "see" the impossibly thin line of either the major axis of the ellipse or the minor axis. The photon is an incredibly thin pancake. Following the argument from Part 1, the photon is also getting bigger (expanding), so in one dimension it is growing enormously compared to the other two dimensions.

In the currently accepted view of the universe, it is known that the negatively charged particle with which we are familiar, the electron, is of much smaller mass than the familiar positively charged particle, the proton, although there exists the electron's opposing anti-particle, the positron, and the proton's opposing anti-particle the negatron (or anti-proton). However, there appear to be far more electrons and protons than there are positrons and anti-protons. In general, matter seems to be composed mainly of combinations of electrons and protons and there is a marked dearth of antimatter comprised of positrons and negatrons (anti-protons). It may be that inside the photon the t+rd and t-rd are the same mass, or the t+rd is much more massive as in normal matter, or the t-rd is more massive - which might explain the preponderance of the less-massive electron and more-massive proton in general, a sort of cosmic balancing-up. It would seem that of the three possibilities (same mass, more massive t+rd, or more-massive t-rd), only one provides the symmetry which would be the result of exactly equal masses or energies. In my estimation, it would be the same sort of odds as getting a perfectly circular planetary orbit, or a parabolic one, rather than an ellipse or hyperbola for a planet/star interaction, i.e. the planets in the solar system and most comets have elliptical orbits round the Sun, but occasional objects come flying through the Solar System, get deflected by the Sun, then fly out of the Solar System - we say they have "escape velocity" and their path is a hyperbola. So let it be with photons.

This provides an interesting scenario for the photon, whatever the orbital path of its components. To us, it vibrates in a plane as the tirds rush round each other, and due to relativistic effects the mass of the components increases with increasing mutual velocity. Continuing with the planet/Sun analogy, as the less-massive tird reaches aphelion it is travelling at its slowest speed relative to the other tird. It consists almost entirely of energy and is massless (almost) so the attraction of the positive and negative charges on each particle causes maximum acceleration due to negligible mass despite the distance between the particles being at its greatest. However, as it swings round the more massive tird at perihelion, it is going at its fastest near light-speed, and its mass is at its maximum. The consequence is that the total mass of the photon is increased at the expense of the energy of the speeding tirds, and is now located between the tirds at their closest point. At aphelion, the total mass is less and is located on the opposite side of the more massive tird. The photon vibrates and its mass changes slightly in time with this vibration - what we know as its frequency.

This brings us to conveniently to the next chapter.

6.

Movement

Pass the Parcel......

In the previous chapters one of the implicit assumptions was that objects in our welkin moved, but could not exceed the speed of light. We can now look at the actual mechanics of movement, what it actually is and how objects achieve it.

Summing up the previous assumptions, objects in our welkin are expanding uniformly, and are comprised of tirds as per the photon described in the previous chapter. Consider the following diagram of the disposition of mass in the photon at different points in its cycle. For the sake of simplicity, we will consider the t-rd as more massive than the t+rd (notice I do not use the terms "heavier" and "lighter", which have gravitic implications).

The X represents the centre of mass of the photon, closer to the more-massive t-rd than to the t+rd. The value of the velocity of the t+rd relative to the t-rd is at its smallest value.

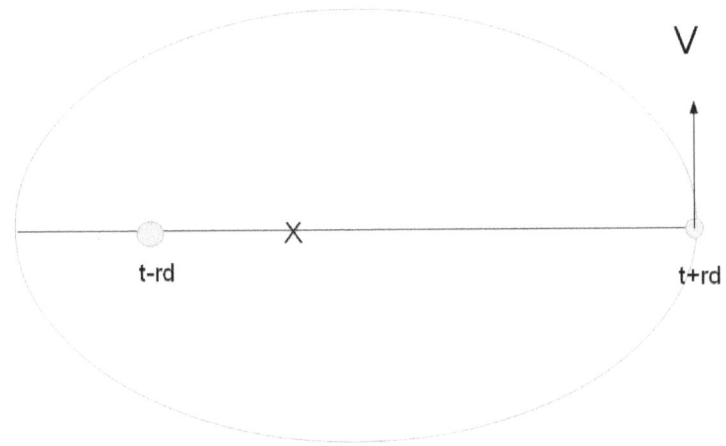

Consider the next diagram, when the tirds are at their closest.

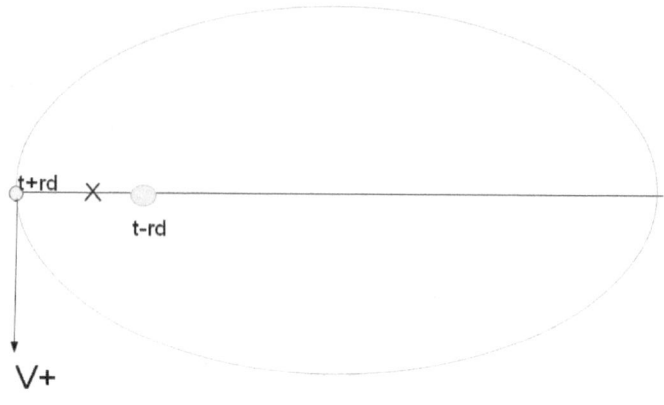

Again, the X indicates the centre of mass of the photon, the tirds are at their closest, the velocity of t+rd is now V+, its greatest value, and approaching the speed of light. The mass of t+rd is increased by relativistic effects. The centre of gravity is PROPORTIONATELY nearer to the t+rd and further from the t-rd. This would simply make the photon vibrate in a continuous-time-universe, but in our welkin where time is quantum and the size of the photon is increasing, should there be any synchronicity between the quanta of time and the position of the tirds in their ellipse, the effect will be the displacement of the centre of mass mainly along the long axis of the ellipse. The progress of the photon will be in minuscule bursts with a wobble caused by the same effect along the minor axis.

The effect to us will be a wave propagating through space with a frequency determined by the size of the ellipse and amplitude determined by the minor axis. How fast is it going? Almost light speed, where light speed is the fastest possible speed of anything in our universe, any faster and the photon/particle would break up as the tird orbit becomes hyperbolic rather than elliptical, and the photon ceases to exist. So, is there any synchronicity between the quanta of time and the position of the tirds? Of course there is, didn't you read Part 1?

A beam of light/radiation consists of a string of photon pancakes with their minor axes pointing in random directions. Of course you can separate out the photons by stopping the ones with minor axes pointing, say, north and south, and just let the east west aligned ones through - and get polarised light/radiation.

7.

Mass Migration

All together now.....

What about larger particles? Objects in our universe, consisting of billions of particles made from our building blocks in various combinations, seem to be either stationary or whizzing about at high speeds relative to us, and in our experience all require force to be applied to them to get them moving or change their direction of motion. We see that the change of motion, or acceleration, is proportional to the force applied but inversely proportional to the mass of the object. Why is it that, once moving, they don't stop (forget friction for the moment)? Somehow, the applied force is converted to momentum and conserved as kinetic energy, and as far as we know, never fades away or radiates away, only being converted back to radiation energy or temperature rise in collisions with other objects. It would be interesting to now the mechanism that converts force to motion, so here is my theory.

All matter is comprised of smaller particles crammed together and held in place by a combination of the expansion of the particles, electric charges emanating from local differences in the positions of the tirds, and magnetism generated by the particles as the tirds whiz around each other making an electric current. Depending on the combinations of tirds making up the electrons/protons/etc., different atoms are constructed which clump together to form solid matter. I say solid matter, because without energy the temperature of the matter will approach absolute zero.

However, as atoms formed by elementary particles have no reason to be stationary relative to each other, various collisions with other atoms and particles create heat and the temperature of the participants in a collision rises. Some collisions cause the atoms to stick together and create molecules. In any case, heat is created by the conversion of kinetic energy to a temperature rise in the atom which manifests by a greater vibration of the fundamental particles, i.e. an increase in the amplitude of vibration of the fundamental particles. Should this vibration become sufficiently violent, it will break the cohesive forces which bind the solid together and thus reduce the solid to a liquid or gas. Given sufficient heat, the vibration will eventually destroy the atom itself. We need to look closely at the structure of the atom to understand what is happening at the fundamental particle level.

Here's a diagram of the smallest bits involved in the make-up of an atomic particle:-

For simplicity, these minute particles are arranged with major axes north-south or east-west, but in reality the axes point in random directions three-dimensionally. The diagram represents a little piece in the heart of a proton. Each of these particles is expanding and its centre of mass is moving in the direction of that part of the ellipse where the greatest mass is present.

Now, if a force is applied to an object, we know it accelerates until the force ceases whereupon it continues to move at a constant velocity until another force is applied. The applied force causes the fundamental particles to turn so that the long axis and greatest mass migrates in the same direction as the force is applied. As the object expands, it does so more in that direction than any other - to us, the observers, it moves! The more force applied, the more the fundamental particles are aligned and the greater the velocity of the object, until all the fundamental particles are aligned exactly the same way, after which no matter how much force is applied the object can go no faster, it has attained light speed. It has also become infinitely long and vanishingly thin - i.e. it is radiation. Obviously, any cohesive forces which bound the atoms of the object together have been overcome as the fundamental particles are now photons (or whatever you want to call them).

The force applied to the object is therefore utilised to change the axis of rotation of the zillions of tiny gyroscopes which are the fundamental particles of the object. The momentum of the object is therefore provided by the angular momentum of these tiny gyroscopes, thereby conserving total energy. In practice, no force can be applied to an object which increases its velocity as it approaches light speed - how can the force "catch up" to it from behind if it is travelling at light speed? Only forces which divert or slow it down can interact with the object.

An interesting side effect of velocity compared with the observer (us) is that, while an object once at rest (relative to us) is travelling at or near relativistic speeds, it is still expanding but also elongating in the direction of travel at the expense of its width and depth. Whatever mechanism is used to slow it down or stop it (relative to us) will result in a restorative change of shape back to "normal" rest shape, although of course larger so that to us, the "expanded" observer, it is the same size. This deformation of shape should result in energy being exchanged between kinetic (in the form of angular momentum) and temperature of the object (it should heat up), the energy being converted from kinetic to heat (vibration of the fundamental particles). Theoretically, then, the object should be hotter and SMALLER as far as we are concerned.

As part of the object's expansion is being used for movement, the overall expansion rate would be diminished, therefore time itself would be slowed down for the object (from our point of view, though not for the object itself). Intriguing thought, would star travellers come back to Earth as "little people"? Anyone building a spaceship to travel to the nearest star to Sol with an inhabitable planet would make sure that it accelerated at roughly 10 metres per second per second for the first half of the journey, turn over at half-way and decelerate similarly for the second half, thus preserving "gravity" to support any human life aboard long-term. As the Bible recounts several people going off to visit "heaven" and returning, and their ages seem to vary between 500 and 900 years (Earth years elapsed, not their own perceived age), it's easy to draw a couple of spheres of smallest and largest distances for the target star, and work out average "speed" relative to Earth for the journey. I wonder if those visitors to another world came back smaller than when they went, and how many of them could stand on a pin head? Sorry, I am being facetious, but it would be interesting to find out if there was any account of size reduction for some Biblical figures. Would their body chemistry be affected? Could they still eat normal food?

Well, enough of this conjecture, it's time to move on to the time part.

Part 3

Time Out

8.

All the Welkin's a Stage

Bit Players......

It is commonly accepted by modern theorists of all persuasions that time itself is relative to the observer, or to put it more poetically, time is in the eye of the beholder. Individually, we all experience time passing at apparently different rates, sometimes slowly, sometimes it seems to whiz past, but eventually there is never enough. We turn to various mechanisms to agree on a standard time, the ticking of a clock driven by a spring or pendulum, the vibration of a crystal, atoms decaying, observations of heavenly bodies, and we all agree that a second is a second, a minute is a minute, and a day is a day - hang on a minute! (pardon the pun) - for someone living on, say, Mercury, a day lasts almost forever! Someone living on Earth a few million years ago would say a day was 23 hours, or 22, or less depending on how far back in time you go. The Bible says that God created the world in seven days, but how long was a day to God? The world hadn't yet been created, so he wouldn't have been thinking about 24 hours..... Perhaps a billion of our present years would be only a day to God.

So, we must agree that time is indeed relative to the observer, and we know that events take place in zillionths of a second in the sub-atomic world, yet take immense amounts of time in the universe - how long does a black hole exist, for instance, very small ones vanish in microseconds but large ones evaporate exceedingly slowly (according to modern theories). For a black hole, perhaps they all live exactly the same length of (subjective) time.

To cut across all of this, I propose to call our time, which you and I experience during our existence, by the name "time", which is very convenient as that is what we call it anyway. But I am also going to give a name to an underlying sort of "time" which I will call "bigtime". Things take place in bigtime in a different order to events we experience in our time.

In previous chapters, we made the assumption that time is quantum, in other words it occurs in extremely brief packets. We also made the point that events do not necessarily take place in time but time is actually governed by the passage of events. It follows that the ticking of a clock, be it clockwork, pendulum, atomic decay, whatever, is not being governed by time, but time is being governed by the clocks (events). The current scientific definition of a "second" is actually a number of transitions of a caesium 133 atom (see [Definition of Hz in Wiki](#)), so we already accept that our time is actually a series of events rather than a constantly-flowing stream which dictates the series of events.

"What difference does it make?", you may ask. Well, it means that we can take time out of the laws of physics just as we can remove gravity, and state that the laws of physics define time rather than time defining the laws of physics.

If events define time, it means that, as all events involve the movement of objects or radiation, then time depends on the position of the bits and pieces that make up our welkin.

9.

Blueprints, Plans and Maps

Time to take your physics.....

When you look up at the night sky, whichever direction you choose, you might see stars, you will certainly "see" galaxies, and most definitely you are looking directly at an immense series of black holes. Given an infinite welkin, your line of sight must intersect an infinite number of all three objects. You are therefore seeing a great sea of black holes floating about leaking tirds. A sobering thought. We are adrift in a sea of tirds, none of which we can see or otherwise interact with, as they are not matter and therefore do not impact upon our welkin. Invisible, undetectable, and probably moving completely at random except when they interact with each other, form matter, and appear in our welkin. Presumably, this interaction is more likely to take place wherever there are greater concentrations of tirds, which must occur by chance every now and then in a random movement of tirds. Thus , our welkin is probably a brief nexus of tirds occurring in a random whorl happening in an immense cloud at some bigtime moment.

How does this manifest in our time in our welkin? The proposition is that our time based on events is driven by positions of these tirds such that, obeying the laws of physics which govern our welkin, a path is chosen through the tirds so that the universe/welkin we are familiar with moves along this path and continues to exist in "time" so long as the laws of physics are enforced. The positions of these untold zillions of tirds move randomly and our welkin continuously (in bigtime) moves from one set of positions to another so that, in our welkin, we observe the laws of physics to be obeyed, whereas what we have done in our time is simply move to the next position of the tirds which conform to the laws of physics. This could be an immense length of bigtime later, but in our time, only a quantum of time has passed.

So, what we are saying is that at any instant in our time, we are manifested by the current location of tirds in bigtime, and the next instance in our time is manifested by the positions of tirds in bigtime which conforms to the laws of physics in our welkin. The two instants in our time separated by a quantum of our time are in fact two instants in bigtime separated by any amount of bigtime. As you can imagine, the amount of bigtime which passes between the point where the locations of the tirds were compliant with a moment in time for us and the point when the tirds were compliant with the next of our moments in time according to our laws of physics is probably immense, but NOT INFINITE even though zillions of tirds are involved in every tiny bit of space in our welkin.

How does this fit in with our experience in our welkin? Like this:-

The configuration of the tirds at a moment in bigtime constructs the welkin we perceive, which is static. The position of the tirds changes in bigtime and conforms to our next moment of time. Our engines of cognizance are re-assembled. We perceive the welkin. A quantum of time has passed.

How can it be that all the zillions and zillions of tirds in the immense sea of bigtime, even after an almost infinite amount of bigtime, are assembled into a configuration we require for our next moment in time? The answer is, they don't have to be in exactly the positions dictated by our laws of physics. We exist in what might be defined as a macro universe, and as long as we perceive it to be our universe at that macro level, it does not have to be exact at the micro level. Thus, we move to the next quantum of time in accordance with our laws of physics at the macro level but if we look very closely at the matter in our welkin, the smaller we are able to perceive, the less likely is it that things will be as they should be according to our laws of physics. The position of smaller particles is uncertain, but the larger structures observe the laws in a sort of generalisation. Heisenberg correctly identified this situation with his Uncertainty Principle. The deeper we look into the microcosm, the wilder the discrepancy between what we see and what the laws of physics say we should see.

Our entire welkin is therefore constructed from a path through the tirds in bigtime dictated by our laws of physics, and a small change to any one of those laws will result in a totally different welkin, i.e. a parallel universe. Except it is not really "parallel" as it might never cross our paths in bigtime. Obviously, there is an (almost?) infinite number of such welkins each following its own path through the tird sea - any combination of physical laws giving rise to a viable configuration for life to exist in a welkin could create another version of you and me living almost identical lives in a "parallel" universe. Or maybe there is only one such combination of physical laws which can do this, there is only one Creation and we are in it. What it does not mean is that, although every particle can occupy every location at one bigtime or another, the postulation that "Anything that can happen does happen" does not hold true, as our welkin time can only exist according to the laws of physics, thus cause and effect are maintained by our path through bigtime and any divergence from it results in a different welkin. As far as we are concerned, there can be only one.

10.

Truth, Dare, or Consequence

Free Will and Witchcraft .....

What happens if we make a decision which affects our welkin? It could be a monumental decision, say, to drop a hydrogen bomb (a massive decision for our tiny, tiny bit of our welkin, but insignificant on even a planetary scale long term), or to sprinkle some salt on our dinner. Is that decision pre-ordained and therefore unchangeable, and we are all prisoners of an inexorable fate? Or does our decision really affect our path through bigtime so that the next quantum of time is changed and therefore our welkin pursues an ever-increasingly divergent path through bigtime? Did that third-class passenger who threw his cigarette end off the back of the Titanic as it left Queenstown change the course and speed of the ship insignificantly but sufficiently enough for it to meet its fateful assignation with the iceberg, when otherwise it would have just missed it or hit it a slighter or greater blow which would not have ruptured enough watertight compartments for Titanic to sink? How slight an action or decision does it take to change the path of our welkin?

There are many known instances where, by the act of observing an event, the outcome of the event is changed. Several instances are demonstrated by experiments wherein a change effected by us on an object causes a corresponding change to take place to another object where, in our welkin, there is no conceivable way the two objects can interact. One of many such instances is Quantum Entanglement (see [Quantum Entanglement in Wiki](#)) where once-closely-related objects continue their association long after being separated. On a larger scale, it can be argued that extremely distant galaxies only spring into existence when they are viewed by an observer, but this cannot be "proven" by any means of which I am aware. Certainly, there is no evidence of a "gravity wave" attendant on the creation or destruction of a galaxy or two. If there was, we could probably find a way of "surfboarding" it to a nearby star.... dream on!

None of the currently accepted models of the universe explain these phenomena, and the convolutions of ten, eleven or twelve dimension string theory seem to do little to help, although I suppose you could consider that entire dimensions of the "real" universe are wrapped up in coils inside particles of our universe. However, there seems to be no evidence that gravity is a weak force because it manifests in more dimensions than, say, the strong nuclear force, and thereby dilutes its power outside our three-dimensional universe - but surely this sort of conjecture has no more "proof" than the theories I am putting to you in this tractate, wherein individual minor actions/decisions influence the future path of our welkin to sometimes enormous effect and sometimes no discernible effect whatsoever.

There are many other difficult-to-explain features of our welkin, some of which veer into areas usually dismissed as fanciful, e.g. the occult, seeing into the future, re-incarnation, and so on. True, some of these are quite simply fantasies or delusions, but there remains a substantial body of evidence that show events can sometimes be dictated by thought or that thought can be transferred telepathically. If enough people believe something to be true, especially on an atomic scale, it becomes true when people who expect to see a particular result actually observe it. Does this mean that belief in something can influence our path through the tird cloud to make it so? Does steadfast belief in a religion actually create the conditions which make the religion "true"? Does belief in God create God? Or did God design the rules of physics to map our path through bigtime in the first place? I cannot put any theory forward on these philosophical questions, but I can certainly see that acceptance of my theories throws an entirely different light on many things. When a coven of witches meets and casts a spell, does it influence our welkin and bend it towards their desired outcome? I suspect that, at times in the past when scientific certitude was weak, groups of consciences might have influenced the welkin, and magic could have been possible. It may still be possible when groups of believers are gathered together, but when some "magic" or otherwise non-logical event is conjured for the public at large, the field of doubt and disbelief generated by sceptical observers overwhelms the desired divergence from the expected normal path of the welkin, and the expected event/ghost/spirit/levitation fails to materialise.

There are some aspects of this which yield surprising results. Given that a moment in our time is in fact a moment in bigtime when all particles are in a certain position, then it follows that all moments in time actually last forever and could be "revisited" by our time, just as all possible moments in time exist in bigtime but may or may not be visited in our time depending on the path chosen by our welkin. Is there any mechanism which prevents our time revisiting a particular configuration of bits in bigtime? Well, there is in the sense that our welkin is a constantly expanding one (or shrinking - as unlikely to be static in size as the chance that two particles orbiting each other make perfect circles or a parabola rather than an ellipse or hyperbola) so our path dictated by the laws of physics prevents revisiting a particular set of particle locations, but there is nothing to prevent a visit to a similar pattern, just on a larger (or smaller) scale. When our welkin guided by the laws of physics moves to the next roughly-approximate (on a micro scale, but exact on our macro scale) set of laws-of-physics-conforming positions of the particles in bigtime, there would seem to be a very good chance that a moment in our time is indistinguishable (to us) from another moment in our time - how would we know the difference?

Obviously we could tell by comparing physical things, clocks for instance, which by the laws of physics must have moved on (that's how our path was chosen) but, our observer engine of cognizance being recreated, we perceive our welkin as a jumble of patterns which our consciousness possibly interprets differently for each of us - i.e. our interpretation of what we see is subjective, just as a colour-blind person "sees" a different picture to a non-colour blind person. A set of patterns once interpreted at one moment in time must predispose us to interpreting a similar set of patterns in the same way, i.e. we recognise them. So, for each of us in our welkin our consciousness is interpreting common patterns we perceive in ways we may mutually agree may be certain things but there are probably many things each of us "see" differently.

11.

Breaking the Mould

The Philosophers Stone .....

We are now at a point where the whole monolithic structure of the universe that was familiar to you before you started reading this book must be crumbling, as any one of the theories in the preceding chapters corrupts the fabric from which that universe was constructed. You can now look round and recognise a fundamentally different welkin from the one you previously thought you inhabited, a bit like those eminent clergymen who once believed the earth was flat but have now looked through Galileo's telescope. So, how real is the world you see? In a physical sense, very real indeed. If you bang your head on the wall, it hurts, even though you know that, in reality, the wall and your head are comprised of atoms which are 99.9999999999999% empty space. If you slip, "gravity" will smash you down against the Earth. The universe is still made of immutable atoms which can only be changed by nuclear reactions. And so on.

However, there is a vanishingly-small chance that when you bang your head against the wall, your head will slip through the wall as if it wasn't there (don't try this at home!). There is no chance that "gravity" as postulated by accepted theories will not smash you down if you slip, but if my theory is correct (no gravity) there is again a chance, though vanishingly small, that you might slip through the Earth (well, perhaps just part way through it... nasty!). And those immutable atoms - well, according to my theory, they are being created anew with the passage of every quantum of time, re-assembled for us in bigtime, according to our laws of physics.

Now, we have already discussed what the result would be if only one tiny aspect of any of the laws of physics were to be changed - we would be in a different welkin. What would happen if we could just change one little law temporarily and then revert back to "normal". Could this be done for a small local volume without affecting the big picture? Is this how witchcraft works, if it works at all?

In the past, there was considerable belief in the existence of a Philosopher's Stone, which, amongst other things, had the power to transmute base metals into gold. There is an attractive explanation for modern man to believe this was a fantasy engendered by man's greed or wishful thinking, but an awful lot of clever people had faith in the existence of this fabulous device. There have been many scientific explanations for this belief, for example it is possible the ancient Egyptians used electric current generated by sticking a copper nail and an iron nail into an acidic fruit thus producing an electric current to electroplate some artefacts made of base metals with gold. This could have given rise to the Philosopher's Stone legend. But supposing that one or more people had sufficient strength of mind, possibly using some means of focussing their thought (such as a Philosopher's Stone), as to be able to briefly overcome a law of physics in their immediate locality - there are many stories of people with the ability for "scrying" using a stone or a crystal ball.

I once read about a device created by the Germans during the Second World War which consisted of a stainless steel disc incorporating chambers filled with mercury, which could be spun as a gyroscope at very high speeds, thus breaking some laws of magnetism (in theory). The intention I believe was to generate some form of anti-gravity or propulsion, but the effect was to destroy the fabric of the universe in a small locality, causing disruption of objects and transmutation of elements in an unplanned way. I do not know if this account was true or an invention, and I have been unable to trace the source again.

You may wonder why this chapter has diverged

somewhat from the mainstream theme of this book, a proposition that the currently accepted theories of our universe are incorrect, onto topics such as witchcraft, Philosopher's Stones, fanciful Nazi artefacts, and the like. Well, the point is this - it would not take a great deal of money, time and effort (compared with, say, building a Large Hadron Collider), to build an experiment which breaks the laws of physics as currently understood. If I am right, the effects should be startling to say the least. (Do we really want to do it, just as the end of the Mayan Calendar approaches and according to which the possible end of the universe is nigh?). It would prove, one way or another, that these theories are correct, and our welkin is not the universe the scientific establishment say it is. I am confident that several of you readers can see a method of doing this and have the means to set such a project in motion.

Part 4

Any Dog

12.

The Right Path

Paved With Good Intentions...

At this point, there are two eventualities. Either you think there is something in the theories put forward in this book, or you dismiss them ALL as nonsense (see first line of preface). In the latter case, none of the rest of this book means anything, and my advice is to throw it away immediately (what do I care, I've got your money...) and forget about it. If you are still reading, you must at least half-believe there might be some shred of truth in all this. A further two possibilities then arise. Either some or all of these theories are right, or none of them are. Again, in the latter case, little harm done, you and I can congratulate ourselves on keeping an open mind and considering alternatives to the official line, life will go on much as before, unless of course a great number of readers are "converted" to this illogical view of the universe and the science of physics is thereby set back a hundred or so years.

But if any of these theories are right, we are looking at a fundamentally different welkin to the one we grew up in, and most of the cosmology and atomic physics on which are built the cathedrals of academe must crumble and give way to new models.

I now feel pangs of anxiety, did I really want to do this? The comfortable universe we inhabit, where science seemed to have cracked most of the problems, will fall about our ears, scientific chaos will ensue, and discomfort and change will become the norm. I put down my ideas more or less as a protest against the refusal of the establishment (any establishment) to consider anything not in their own self-interest, with the best of intentions, but there is an old proverb that starts "The road to Hell is paved with good intentions..." and perhaps I am supplying some paving slabs. Simply by writing this book, am I diverting the path of our welkin away from a cosy future, and figuratively setting the course of Titanic for a meeting with an iceberg? On the whole, there is something to be said for accepting everything that happens as inescapable fate, pre-ordained from when time began (our time, that is), and there is nothing we can do about it. However, I think I prefer to believe that I am helping to pick an improving path through bigtime by my actions, and I hope you do too.

13.

Corroborative Evidence

Some Explanations...

If the theory that everything in our welkin is expanding is correct, there should be several things which the theory explains which cannot be explained by currently accepted theories. There are several candidates for such illustrations, and they range from simple everyday things you can observe to logical arguments about, say, the location of sub-atomic particles.

Let's start with surface tension and capillarity. You can observe this wherever a liquid is in contact with the surface of a solid. It can vary from the well-defined meniscus of a drop of mercury on a plate which maintains the shape of the blob, to the spreading action of petrol dropped on a paving stone when it spreads all over. Generally these two states are labelled capillary repulsion (when the force of attraction between the molecules of the liquid to each other is greater than the force between the molecules of the liquid and the molecules of the solid) and capillary attraction (when the force of attraction between the molecules of the liquid to each other is less than the force between the molecules of the liquid and the molecules of the solid). What is this force? It is not electrostatic. It is not magnetic. It is not gravity (mass attraction - which does not exist anyway according to Part 1). So what is it?

Of course, it is simply that the molecules of both the liquid and the solid are expanding. In forming any solid or liquid object, the molecules interact with each other so that the shape of the solid is maintained and the volume of the liquid remains constant. In a solid, they are arranged in lattices or patterns to form crystals or similar with standard patterns of alignment depending on the particular state of the object, especially in terms of its temperature - the alignment is destroyed or becomes looser when temperatures rise due to the vibrations of the particles which is the manifestation of heat. We have already discussed the way in which objects move by expansion, biased along the major axis of the ellipse formed by the orbiting particles within each fundamental building block.

When a liquid is in contact with a solid, the crystalline molecular structure of the solid remains constant but the molecular structure of the liquid is constantly changing (otherwise it becomes a solid).

At the boundary between the two substances, the molecules of the liquid are in interaction with the molecules of the solid which causes the boundary molecules of the liquid to predominantly align with the surface molecules of the solid. The liquid molecules, unbound by attachment in matrices with the other molecules of the liquid, expand predominantly along their major axes thus creating an imbalance between the expansion rate of the solid and the expansion rate of the contact surface molecules of the liquid and therefore a current in the liquid as the molecules move relative to the molecules of the solid. As the volume of the liquid is maintained, the surface molecules of the liquid are constantly replaced in position by interior molecules and the alignment randomised as a result. This can be easily verified by an examination of the movement of molecules at the boundary between the liquid and the solid, thus proving the theory and explaining why some liquids move up a capillary tube until an equal and opposing force (not gravity!) cancels out the surface tension capillary attraction... Unless you have a better idea of what causes it? Please don't cite surface tension, that is an effect, not a cause!

Moving on from the fairly easily observed phenomena to the other extreme, current theory implies that a particle moving from one point in the universe to another passes through every point in the universe on its way. This is very difficult to explain in the normal view of the universe, it seems contrary to all logic if not absolutely impossible. But if we view it from our welkin, adrift as we are in bigtime, there's a good chance that any tirds have been just about everywhere at least once before a pattern occurs in bigtime which allows a quantum jump according to our roadmap (the laws of physics) and the next quantum of time in our own welkin wherein the macro world we see conforms to expectations and we can exist, i.e. are re-created.

However, there is no guarantee that a particular particle has visited a particular location, as the bigtime elapsed is not infinite in bigtime and even if the bigtime elapsed is immense there is a possibility that the particle did not visit a particular location. Which means that, as we cannot detect how much bigtime has elapsed between two quanta of our time (or even if they are in any sort of sequence in bigtime at all, if that is a meaningful statement in bigtime), the chance of it happening could be completely random. Even if we devise some means of measuring elapsed bigtime between two quanta of our time, the appearance or not of a particle at any particular point between quanta is random, we can only know that macro particles have been re-created at the point predicted by the laws of physics as expected in our time because OUR TIME IS DEFINED BY THIS EVENT in bigtime. Meanwhile, (that's bigtime meanwhile) the chances must be zillions to one that every particle has visited every location between quanta of our time when a required particular pattern of particles occurs, considering the immense number of particles involved in forming that pattern.

Unless of course there is some overarching force or condition in bigtime which coerces the particles to form certain patterns or at least influences them to favour certain configurations. This could be as rigorous as a mould shaping molten iron poured into it, or as gentle as a breeze favouring sand dune patterns in a desert.

My mind favours a sort of tendency for recurring patterns to occur, much as when rain forms a puddle which depresses the ground so that subsequent rain tends to form a similar puddle in the future. Eddies in the bigtime particles might cause whole series of similar patterns to repeat, each pattern reinforcing a tendency for the next pattern to be similar, or for a sort of chain reaction of patterns formed one after another.

I am constrained to discontinue this train of thought, as, whilst I am quite happy to conjecture about the theories creating our welkin in bigtime, I must draw the line regarding conjecture of what causes events in bigtime (does it exist in a sort of "supertime"?) - this is a step too far for me, as I can see it continuing beyond supertime ad infinitum. Unless of course "supertime" is actually our time.... the mind boggles.

14.

When the Bigtimes Come

Captious Sands.......

Stepping back to the relatively simpler concept of the randomly-swirling cloud of tirds, the question must arise, "Are all tirds the same?", apart from being either a t-rd or a t+rd, that is. Well, if they are to be considered the fundamental building blocks they have to be the same as far as we are concerned. In which case, is it possible that there is only one t-rd and one t+rd, just whizzing about a lot?

A similar proposition was made about electrons actually being all just one particle, the charge on it being either positive of negative as it moved through time either forwards or backwards ----
(see <u>Every Electron is the Same Electron</u>) by famous physicist John Wheeler back in 1940, but it is rumoured he

made the suggestion in jest just to provoke a discussion about it.

In my interpretation of the same question regarding tirds, I believe that there may be only one original tird and that all the others may be virtual tirds created by its presence. One positive tird in one spot (a single t+rd) would immediately invoke a surrounding cloud of negative tirds (t-rds) each of which would itself invoke a surrounding cloud of t+rds, and so on. Does bigtime consist of a cloud of virtual tirds instantaneously brought into existence by the creation of a single original tird? Could there be just one single particle which, by quantum effect, spontaneously splits occasionally or cyclically into a positive and negative tird each giving rise to the bigtime tird sea of virtual tirds? Could it just have happened once, and the continuing reverberation and echo of it actually be bigtime?

In all of these cases, every tird is in effect an image of the original tird, i.e. is actually a manifestation of the same thing. Even so, how can an effect on any one of the instances of a tird be reflected on another instance of the tird somewhere else? Does the effect of an event on one virtual tird get propagated back through the tird sea to the original tird instantaneously, and then back out to its images? If so, all tirds should be affected, but we observe the effect only on two particles once associated with each other, so this cannot be an explanation. No, there must be a continuing association between two once-associated tirds in bigtime. I suspect it is to do with some property of electrostatic force, which is tied in with the concept of the fundamental particles forming a black hole (see Part 1) in our time which endures in bigtime so that the separated particles (in our time) are still one object in bigtime. Could electrostatic attraction be in the form of a sort of elastic tendril rather than a field? Ridiculous thought, how could it? Although sillier ideas have turned out to be true.

Could it be that all t+rds are connected inseparably to their simultaneously-formed t-rd, so that the manifestation of a t+rd and a t-rd in our welkin are actually the two ends of a string? Is all matter composed of these strings, break the string and the tirds fly apart, become non-matter, leave our welkin? So, if all matter is made up of tird "strings" can the two ends be used to form different objects/particles? When the particles are separated, does the string "break" if the distance becomes sufficiently great, or not?

At this point, I will change the word I am using to connect the t+rd and t-rd from "string" to "chain" so as not to confuse things with "String Theory" ---
(see [Wiki entry for String Theory](#)), although obviously there are similarities - e.g. everything is made up of strings. We'll call it "Chain Theory" and say that, whilst the manifestation of matter in our welkin is the t-rd and t+rd at each end of a chain, there may be any number of "intermediate" pairs of

tirds between them. In fact, as the distance between the tirds at each end of the chain increases, it may be that new pairs of tirds spring into existence to "fill in the gaps" where the chain is stretched thin. So, an event which impacts on one end of the chain travels down the chain and causes an effect at the other end. As the intermediate tirds all cancel each other out, they have no effect in our welkin, as far as we are concerned they don't exist as they do not form matter.

Wow, is this a viable explanation for Quantum Entanglement? At least it is better than just looking at the problem and scratching our heads, as current theories have us doing. Stop fiddling with your balls of String Theory and join the Chain Gang!

Maybe it's a flight of fantasy too far, so why don't you come up with your own theory? As there are no better explanations I know of as I write this, you have an open field, and after you have read the next chapter you might be inclined to put pen to paper (or fingers to keyboard) and launch your own version of the universe we live in.

Meanwhile, if we cast our minds back to how the sea of tirds was created from the original tird (see earlier), perhaps the sea is in fact one huge chain created by the separation of the components of the original tird. But every tird in the chain can then create its own chain if it is separated into a t+rd and a t-rd. So we have an infinite number of chains springing from the original chain. Is every one of these chains a separate tird sea? A separate "bigtime"? Is a quantum of time in our welkin a jump from one chain to another or do we stay on the same chain or are all the chains mixed up in one vast bigtime Gordian knot?

Modern String Theory seems to require ten, eleven or twelve dimensions to function. This Chain Theory has either bigtime (itself four-dimensional?) or an infinite number of bigtimes, you may take your pick of which version you prefer.

Well, by now I have led you far away from the conventional view of our universe as defined by accepted theories of physics and promulgated by the scientific establishment. We are wandering about in a wilderness without a solid scientific basis of proven truth, no vantage point on which we can stand to evaluate our situation.

However, all is not lost, and you are ready to read the final chapter.

15.

Conclusion

The First of Many or The Last...

Having destroyed gravity, disrupted time and reduced all matter in the universe to a heap of tirds, where do we go from here? Can't really start from scratch and devise all physics from the ground up, so I guess we will have to look at amending existing accepted equations and formulae to remove references to gravity and time.

Well, removable of gravity should be possible, but time is a different beast. As we live in the stream of time, stepping outside it renders our equations pointless in the welkin in which we exist, as that very existence depends on the ongoing sequence of quanta selected by the laws of physics governing our path through bigtime. As far as I can see, we are faced with an insurmountable obstacle to making any progress in this direction.

Therefore, I fear these theories are the last in their line, no progress can be made along the lines they indicate, and we are stuck with the obviously deficient theories currently in vogue.

Unless you know different.

There is a saying in the hunting/shooting fraternity along the lines "Any dog is better than no dog, but no dog is better than mine!". This can be interpreted several ways, and each axiom derived applied to this dog of a theory.

16.

Addendum on "Gravity Waves".

What did they measure?

Since this book was first published, considerable time, money and effort has been put into detecting "Gravity Waves", mostly with the LIGO and SuperLIGO projects. Eventually, a discernible effect was measured, and attributed to two or more merging black holes.

The effect was detected as the difference in length of two laser paths at right angles to each other, and purported to show the passage of a gravity wave through the laser paths resulting in discrepancies of time taken for light to travel along them. The gravity wave was generated by the destruction of matter occurring when black holes merge to create a single black hole of less total mass then the contributing smaller black holes – the "lost" mass being "radiated" away as gravity waves.

If you have read and understood this book, then you will know that I believe gravity does not exist as a force, and therefore there can be no gravity waves. Please indulge me while I explain why this remains my belief.

Beforehand, I must review some observable facts.

The first of these are the various theories of light red shift, "tired" light, and the universe we inhabit. If light carries on forever, and the universe is infinite, then there is no escaping the fact that the Earth would be burnt to a cinder, as there are an infinite number of stars in every direction and all of them would contribute to the total light reaching Earth. Therefore there must be a limiting factor to the distance light can travel. Observed fact indicates that the further away a light beam emanates from us, the more it is red shifted, and this is usually ascribed to the expansion of the universe, i.e. the further away from us something is, the faster it is moving away from us. Now, it doesn't take a logical genius to see that, if the wavelength of light increases with distance of source from us (red shift) then as that distance approaches a certain point, the source of the light will be moving away from us at (or nearly at) light speed, and we (the Earth) would be the centre of a sphere where all light sources situated on the surface of the sphere are moving away from us at light speed. The wavelength of light reaching us from those sources would be almost infinitely extended. Now, you cannot have a wave with infinite wavelength – it simply isn't a wave! So what do we have? A waveless photon? Nothing? Some sort of particle? Dark matter? Well, whatever it is, it ain't light as we know it.

So, I consider our universe to be a sphere the surface of which recedes from us at light speed. But consider this – the sphere is expanding at an ever-faster rate – the bigger it is, the faster it expands, the faster its surface moves away from us. Light emanating from Earth never reaches that surface, and as that surface accelerates away from us, the photons actually start to drop behind the surface so that, after a time, they are actually moving away from the surface, i.e. back towards Earth. (This could explain why we see multiple images of very distant galaxies, rather than the "gravitic lensing" favoured by many). Now, you will remember that I believe everything in the universe is expanding at the same time (my explanation for gravity), so effectively the surface of Earth will eventually catch up to the photons which emanated from it (if there is no interference from other bodies!). There is no escape from this, and our universe is therefore a black hole by definition and according to observable facts.

Earlier on, I described my belief in what time actually is, i.e. the path thru' chaos determined by the laws of physics which navigate our existence between states of matter (the positions of fundamental particles in our universe), the choice of apparently continuously expanding scenarios as we move from one quantum of time to the next (which we experience subjectively as an always-advancing stream, although in "real" time these positions occur randomly). Time is therefore the exchange of one set of particle positions for another which conforms superficially to positions predicted by the laws of physics even though at microscopic levels this is not true and diverges from these laws at all levels below macroscopic. Time therefore is dictated by the expansion of our universe (and everything in it).

Now, we have already said that the rate of expansion is dictated by the size of the universe (and therefore the particles of which it consists), so time is dependent on the amount of matter (mass) in our universe – the bigger it is, the greater the expansion rate, the faster time passes (objectively, but at the same rate to us subjectively). Alter the amount of mass in the universe, and the rate of passage of time also alters.

When two or more black holes coalesce, current theories indicate that matter is lost (converted into energy as gravitational waves, according to the LIGO enthusiasts). According to my theory, the loss of matter would create a blip in time, propagating thru' our universe as a minute change in the rate of time passage, manifesting itself in the LIGO experiment as a change in the length of "time" for a laser beam to travel between two points at different distances from the black holes event. I would use the term "time quake" but it has already been quoted in science fiction for a different effect.

So, I believe that what LIGO has measured is in fact a perturbation in time itself, not a gravity wave.

You must ask yourself, if you apply Occam's razor to all the available theories explaining our universe, covering string theories with their dozens of dimensions, Einstein's space/time continuum with masses making dimples in it, converging and colliding branes, my theories detailed here, or dozens of other now-defunct (in the view of the scientific establishment) theories, then what is to recommend any particular one? Surely, if you have to invent loads of extraneous things for a theory to work, e.g. Dark Matter, Dark Energy, Gravitons, Gravitational Waves, etc., the theory must be suspect, then logic favours the simplest theory with the least (or no) "inventions", whereas gravity requires these for starters.

Finally, ask yourself this:- If you were in a box with no means of communicating with the outside, how would you know if you were on the surface of Earth or in a spaceship accelerating at about 10 metres per second per second? Is there any way of differentiating gravity from acceleration? If not, why "invent" gravity?

Feel free to pick holes in my logic....

17.

Refutations

A Summary of the main arguments raised against Expansion Theory.

Over the many years since I first proposed Expansion Theory as the basis of the universe we inhabit, there have been objections raised regarding how it can explain various features of reality. Most of these objections were presented with vehemence by opponents/adherents to other theories/dogmatic individuals who refuse to countenance anything new.

The objections fall into four main groups, which can be summed up as:-

1. Orbits without gravity.
2. Holes in solid bodies.
3. Maintenance of relative sizes.
4. Conservation of angular momentum.

17.1 Orbits without gravity.

This is difficult to envision, and so I will try to reduce it to a simple exercise. You will need a sheet of A4 paper, a pencil and a ruler, plus a compass (the sort you draw circles with, as opposed to navigate!). Place the paper landscape-wise in front of you. Measure 8cms in from the left hand edge and draw a vertical line. Measure up from the bottom of the line 5cms and make a small cross, label it E. Measure another 8cms up the line above E and make another cross, label it M. Use the compass to draw a circle, radius 2cms, centre E. Use the compass to draw another circle, radius 1cm, centre M. Draw a horizontal line thru' M. The diagram now represents our Earth/Moon system, though not to scale. For the sake of convenience, bearing in mind that each and every quantum of time is individual and in no way connected to any other quantum except that it has been chosen in sequence by our path thru' the chaos as dictated on a macro level by the laws of physics, we will assume that in underlying reality the rate of expansion of matter between two quanta results in doubling in size. (Obviously, it doesn't, the increase is likely to be tiny, of the order of planck lengths, and time quanta likely how long it takes for a photon to travel a planck length or thereabouts).

Now, where the line between E and M crosses the Earth circle, label point O. The horizontal line thru' M indicates the observed velocity of the Moon by the observer at O. In the next quantum of time, everything has doubled, so draw another circle, centre E, of radius 4cms. Where the line between E and M crosses this circle, label point P. Set the compass to a radius of 16cms, and using E as the centre, draw an arc crossing the horizontal line. Label the point where it crosses N, and draw a circle, radius 2cms around it. Draw a line between N and E. The diagram now represents the Earth/Moon situation during the second quantum of time. To the observer, who has also doubled in size, everything seems the same size as before, but the observer is now at point P and the Moon is now orbiting the Earth at the same observed distance as before, but has moved round in its orbit. Measure a point 8cms above M on the vertical line (at the top of your paper) and draw a line from there to N. This represents the APPARENT velocity of the Moon to the observer at P. By repeating this diagram sufficient times, each time using the APPARENT velocity rather than a horizontal line, you can see that the Moon circles the Earth and that "gravity" is redundant.

Instead of doubling in size between quanta, imagine tiny steps so the motion of the Moon is smooth

17.2   Holes in Solid Bodies.

One of the points raised against expansion is that the expanding matter of, say, the Earth, would "fill-in" the caves and valleys which however persist. I do not propose to go into great lengths over why this should be so, but refer to articles in almost any basic physics book on the topic "Expansion of Holes", which can easily be looked up on the internet.

17.3   Maintenance of Relative Sizes

A persistent argument raised against expansion theory is that, given different effects we call gravity on planets of different sizes and/or densities, how is it that their sizes relative to each other remain unchanged. Obviously, if the gravitic effect is caused by acceleration, then the surface of, say, Earth, must be accelerating at a greater rate than the surface of, say, Mars, and therefore the Earth should grow faster relatively speaking than Mars. This should be particularly apparent in the case of two planets of the same size but of different densities, and therefore of different surface accelerations.

That would be true if we took the particular time of NOW and considered it as the first moment of existence. But of course it is not the first moment, and the surfaces of the planets are already travelling at a velocity determined by previous conditions on that planet. The acceleration of each surface is the change of velocity, not starting from rest as some people seem to assume. Taking two planets A and B of the same size, where A is mostly metals and B is mostly lighter elements, we can use simple mathematics to see that on two successive quanta of time the apparent expansion of both planets is equal (otherwise, they would become of different relative sizes). We can call the "real" distance between a point on the surface in the first quantum and the same point in the second quantum S. Assuming the "gravity" effect on the first planet is double that on the second planet, taking the simple equation of motion

Vsquared = U squared + 2AS

where U is the initial velocity of the surface moving away from the centre of the planet, and V the final velocity, and A is the Acceleration (the "gravity"), we know that A is double for the first planet compared with the second planet, but S has to be the same for both. Then

$(V^2 - U^2)/2A = S$ for both planets.

As A is double for the first planet compared with the second planet, we can say

$(V^2 - U^2)/2*2A$ first planet = $(V^2 - U^2)/2A$ second planet

or $V^2 - U^2$ first planet = $2V^2 - 2U^2$ second planet

thus $V_1^2 - U_1^2 = 2V_2^2 - 2U_2^2$

or in the general case we can say:-

$V_1^2 - U_1^2 = A_1 (V_2^2 - U_2^2)/A_2$

Because we have been indoctrinated in a universe where gravity exists, it is more comfortable to write G instead of A in the equation. Anyway, you can see that by judicious choice of velocities the relative sizes of planets with different densities are maintained, bearing in mind that each and every quantum of time is independent and chosen in sequence in order to conform to our laws of physics at the macro level. In underlying reality, each quantum may be separated by an almost infinite "time" and "place", but physics selects the next quantum for our subjective existence so that the laws are obeyed and the planets maintain their relative sizes and so do all other particles.

17.4   Conservation of Angular Momentum.

This really is a non-starter. In the diagram of 17.1. above the observer at O and "later" at P measures the angular momentum of Earth by its rate of rotation, size and mass. As far as the observer is concerned, these are identical in both quanta of time, and the angular momentum is exactly the same, i.e. remains unchanged.

18.

A Simple Experiment

Well, simple in theory, a bit more difficult in practice.......

In the final analysis, what proof is there in favour of Expansion Theory that does not apply equally to Newton's Gravity or Einstein's dimple in space/time? In another universe where Expansion Theory is the norm, and Gravity is put forward as an alternative, this can be turned round and levelled against Gravity. We are faced with a difficult task, like being asked to measure the dimensions of a building when we have nowhere to stand to take the measurements and no yardstick to measure with. Therefore, we must rely on anomalies of the Gravity theory to disprove it.

There are a number of such anomalies, many of which are well known. For example, the rotation of galaxies does not conform with the predictions of gravitic theory. But the most intriguing one for me was discovered by Edwin Hubble. It is now accepted beyond any reasonable doubt that we live in an expanding Universe. Hubble showed that, as you travel further away in space, everything is accelerating away from us at increasing velocities proportional to the distance away from us. Measurements have produced an average figure for this phenomenon of about 70 kilometers per second per 3 million light years. Now, this holds more-or-less true for, say a billion light years distance. It holds true for a hundred light years distant. How about 10 light years? One light year? A million kilometers? A kilometer? A centimeter? A planck length? Where would you put the boundary where this phenomenon commences? I put it inside every particle in the universe, and that is Expansion Theory.

If my theory is true in every part, you see from the chapter on Movement and Momentum that they are effects of expansion. Matter (all particles) consist of orbiting charges mainly in elliptical orbits, and the orientation of the ellipse long-axes determines the momentum and movement of matter relative to other matter. Consequently, a photon consisting of a small number of constituents (maybe only two, a positive and negative charge) whizzes along at light speed in the direction of its long axis towards its more-massive focus point. More complex matter with ellipses pointing in random directions remains stationary, but commences to move when a force is applied which causes the ellipses, like so many tiny gyroscopes, to change orientation and thus create momentum/movement in the direction of the force (force which can ONLY be produced by interaction with other matter). The change of orientation of the ellipses creating movement is at the expense of the general expansion of the ellipses, so that a moving body is elongated in the direction of travel. A consequence of this should be a general diminution of expansion, i.e. a decrease in size relative to a staionary object.

The experiment I propose then consists of two identical clocks, A and B, next to each other. Clock B is then sent on a journey involving acceleration to a fraction of light speed until eventually returning to be beside clock A. Einstein predicted that clock B would have lost time compared with clock A, and this has indeed been shown to be so in practice. If Expansion Theory is correct in every respect, then clock B should also be SMALLER than clock A in proportion to the time lost.

I look forward to this expeiment being performed, and to seeing the result in the not-too-distant future.

www.ingramcontent.com/pod-product-compliance
Lightning Source LLC
Chambersburg PA
CBHW030949240526
45463CB00016B/2245